Science Tools

J. A. Randolph

Newbridge Educational Publishing,
a Haights Cross Communications Company, New York

Table of Contents

Observation Tools _____ 4

Measuring Tools_____ 6

Tools to Collect and Record Information _____ 11

Special Tools _____ 14

Tools We Use _____ 16

Glossary _____ 19

Index _____ 20

Science Tools
ISBN: 1-56784-415-4

Written by J. A. Randolph
Edited by Gare Thompson Associates
Designed by Silver Editions
Production Manager: Michael Miller

Newbridge Educational Publishing
11 East 26th Street, New York, NY 10010
www.newbridgeonline.com

Photo Credits:
Cover: Ana Esperanza Nance; Page 1: Telegraph Colour Library/FPG International; Page 3: Richard
Heinzen/Superstock; Page 4: (clockwise from top left) Michael Newman/ PhotoEdit, Capece/
Monkmeyer, SCIMAT/Photo Researchers, Inc., Leonard Lessin/FBPA/Photo Researchers, Inc.;
Page 5: (clockwise from top left) Wolfgang Kaehler/Liaison International, Tim Barnwell/Stock
Boston, Stephen Frisch/Stock Boston, John Boval/Photo Researchers, Inc.; Pages 6–7: Nancy
Sheehan/ PhotoEdit; Page 7: (insets) M. Antman/The Image Works; Page 8: Jean-Marc Dessalas/
Gamma Liaison; Page 9: Tom Stewart/The Stock Market; Page 10: Corbis; Page 11: Rich Frishman/
Tony Stone Images; Page 12: Kermani/ Gamma Liaison, Tony Freeman/PhotoEdit (inset); Page 13:
Bill Horsman/Stock Boston, R. Crandall/ The Image Works (inset); Page 14: (from top) NASA/Media
Services, NASA/Media Services; Page 15: (from top) Sylvain Kanamori/Gamma Liaison, Denis
Tapparel/Liaison International; Page 16: Pat Watson/The Image Works; Page 17: (clockwise from
top left) B. Daemmrich/The Image Works, David Young-Wolff/ PhotoEdit, Telegraph Colour
Library/FPG International; Page 18: (tools from top) Will & Deni McIntyre/Photo Researchers, Inc.,
Nikolay Zurek/FPG International, M. Antman/The Image Works, Stephen Frisch/Stock Boston, Index
Stock Photography; (other photos, clockwise from top left) Rosalind Creasy/Peter Arnold, Inc.,
Jan-Peter Lahall/Peter Arnold, Inc., David Young-Wolff/PhotoEdit, Xinhua-Chine Nouvelle/Gamma
Liaison, John Bova/Photo Researchers, Inc.

10 9 8 7 6 5 4

Science is all about observing, asking questions, and finding things out. You can find out a lot by simply using your senses. But you can learn even more with the help of tools.

Scientists use many kinds of tools. Some tools help scientists observe. Why might scientists use tools like the ones shown here?

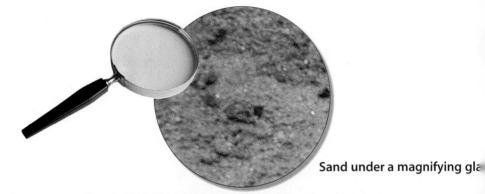

Sand under a magnifying gla

A **magnifying glass** makes objects look bigger. Details are easier to see.

Sand under a microscope

A **microscope** makes objects look much bigger. It is used to study very tiny things.

Bird seen through binoculars

Binoculars make objects that are far away look closer. A scientist can study wildlife without getting too close.

Moon seen through a telescope

A **telescope** helps a scientist see things that are thousands of miles away.

Some tools help scientists measure things. Scientists measure temperature, or how hot or cold things are. A **thermometer** is a tool that measures temperature.

Water freezes at 32 degrees Fahrenheit. Read the thermometers. Which one shows the temperature of the air at a skating pond? How do you know?

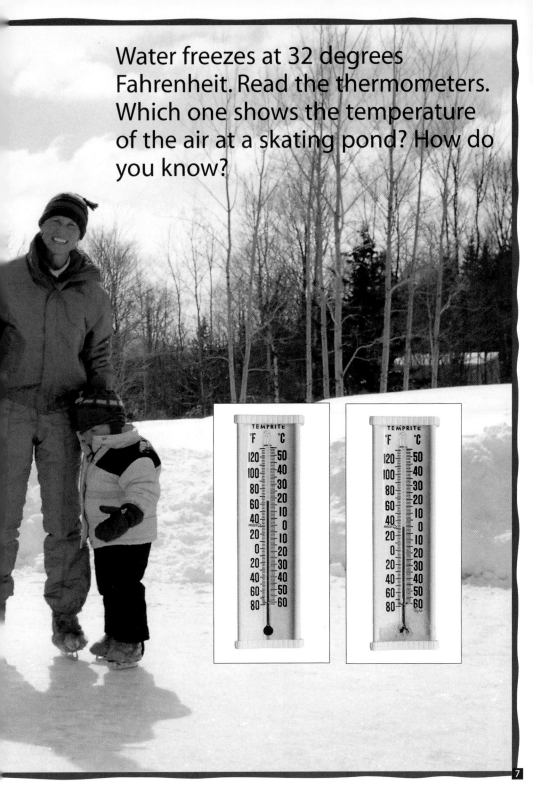

Scientists measure weight, or how heavy things are. A **scale** is a tool that tells how much something weighs.

Doctors are scientists, too. This doctor is weighing a girl. Why do you think this scale is better to weigh a person than the scale on page 8?

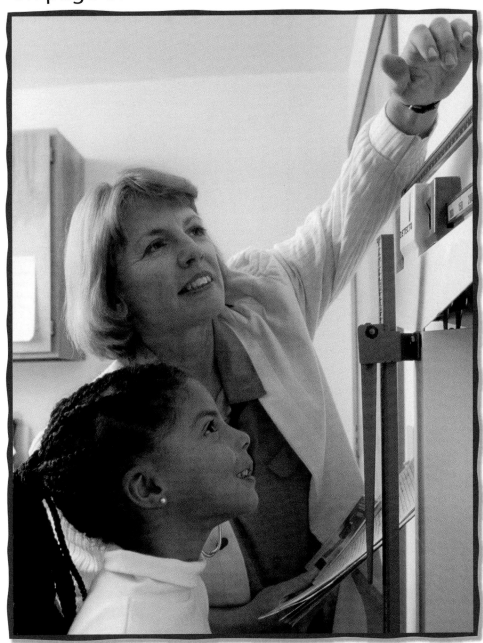

Scientists measure size, or how long, tall, or wide things are. This scientist is using a **measuring tape**, a tool that measures size, to find the diameter of this tree trunk.

This scientist is using a **hammer** and **chisel** to chip rock away from a dinosaur fossil, the bone of a prehistoric animal. The fossil will help him discover information about dinosaurs. Later, he will record his findings.

Scientists use **seismographs** to collect and record information about earthquakes. Whenever the Earth shakes or vibrates, a needle records the information on paper.

Look at this seismograph reading. Do you think there was an earthquake? How can you tell?

Charts, **graphs**, and **diagrams** are also tools. They help make information easy to understand.

This scientist records information about all kinds of weather—from a hurricane's wind speeds to the amount of rain that falls in different regions.

This graph shows the year's rainfall. Can you tell how much rain fell this month compared with last month?

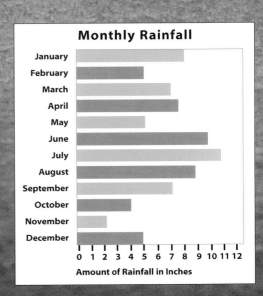

Monthly Rainfall

January
February
March
April
May
June
July
August
September
October
November
December

0 1 2 3 4 5 6 7 8 9 10 11 12

Amount of Rainfall in Inches

Special tools are sometimes needed to explore places scientists could not study without them. **Space stations** help scientists travel into space to conduct experiments.

Space probes help scientists study places in space that are too far away to visit.

How might these special tools help scientists study life in the Antarctic or under the sea? Why do you think they need them?

You need science tools, too. **Paper** and **pencils** are tools you probably use every day.

What tools are these students using?
How are they using them?

Look at the science tools at the left. Which too would you use to observe, measure, or collect information about each object shown?

Glossary

binoculars (buh-NOK-yoo-lurs): An observation tool that makes faraway objects look larger and closer.

chisel (CHIZ-uhl): A metal tool that has a sharpened edge at one end.

Fahrenheit (FAR-uhn-hite): The name of a temperature scale on which the freezing point of water is at 32 degrees and the boiling point of water is at 212 degrees.

fossil (FAH-suhl): The hardened remains or traces of an animal or plant that lived long ago.

hammer (HAM-er): A tool with a heavy metal head on a handle.

magnifying glass (MAG-nuh-fie-ing GLASS): An observation tool that makes small things look larger.

measuring tape (MEH-zhuh-ring TAPE): A long strip of cloth, plastic, or steel marked off in units for measuring.

microscope (MIE-kruh-skope): An observation tool that helps people see tiny things too small to be seen with the naked eye.

scale (SKALE): A tool used to find out how heavy something is.

seismograph (SIZE-muh-graf): An instrument used to measure the power of earthquakes.

telescope (TELL-uh-skope): An observation tool that helps people see things that are very far away, especially in outer space.

thermometer (thur-MOM-i-ter): A tool for measuring temperature.

Index

Antarctic, 15

binoculars, 5

charts, 13

chisel, 11

diagrams, 13

earthquake, 12

Fahrenheit, 6–7

fossil, 11

graphs, 13

hammer, 11

magnifying glass, 3, 4, 17, 18

measuring cup, 18

measuring tape, 10, 18

microscope, 4, 17

Moon, 5, 18

paper, 16

pencils, 16

rainfall, 13

sand, 4

scale, 8–9

sea, 15

seismograph, 12

size, 10, 17

space probes, 14

space stations, 14

telescope, 5, 18

temperature, 6–7

thermometer, 6–7, 18

weight, 8–9